YOUR KNOWLEDGE HAS VALUE

AF143599

- We will publish your bachelor's and
  master's thesis, essays and papers

- Your own eBook and book -
  sold worldwide in all relevant shops

- Earn money with each sale

Upload your text at www.GRIN.com
and publish for free

Isra Bakawasa

The Role of Obesity in Cancer

GRIN Publishing

Bibliographic information published by the German National Library:

The German National Library lists this publication in the National Bibliography; detailed bibliographic data are available on the Internet at http://dnb.dnb.de .

Imprint:

Copyright © 2014 GRIN Verlag GmbH
Print and binding: Books on Demand GmbH, Norderstedt Germany
ISBN: 978-3-656-87915-2

This book at GRIN:

http://www.grin.com/en/e-book/287251/the-role-of-obesity-in-cancer

GRIN - Your knowledge has value

Since its foundation in 1998, GRIN has specialized in publishing academic texts by students, college teachers and other academics as e-book and printed book. The website www.grin.com is an ideal platform for presenting term papers, final papers, scientific essays, dissertations and specialist books.

Visit us on the internet:

http://www.grin.com/

http://www.facebook.com/grincom

http://www.twitter.com/grin_com

Spring | 14

The Role of Obesity in Cancer

Isra F. Bakawasa

M.S. Biotechnology

The Catholic University of America Washington, DC

Tuesday, May 2, 14

Abstract

Obesity is a known cause of cancer, with worldwide research and epidemiological studies supporting the relationship between obesity and cancer. Obesity can lead to a number of cancers, through cellular and metabolic changes in the body. Hyperinsulenimia, low levels of estrogens and Insulin like growth factor binding protein-3 (IGF-BP3) are among the major causes of cancers that result from obesity. In this study we aim at reviewing the metabolic changes that accompany obesity and cancer, the molecular mechanisms associated with obesity and cancer and finally the epidemiological significance of this relationship.

1. Introduction

Obesity is a well established cause of cancer, with worldwide research and epidemiological studies being carried out in order to further appropriate the relationship between obesity and cancer. In order to better understand the role of obesity in cancer we must first fully understand the underlying causes of obesity that lead to changes in biochemistry, physiology and cellular functioning of an individual that aggregate to cancer. There is a need to fully understand the implications of obesity in causing cancer, especially breast cancer in post menopausal women. In this study we aim at reviewing the main causatives of cancer and obesity, the molecular mechanisms that turn obesity into cancer and the epidemiological significance of this relationship.

2. Background

2.1 Obesity

Obesity is defined as an excessive or abnormal accumalation of body fat in dangerous levels that pose a threat to health and quality of life. Being overweight is different from obesity, overweight means that the body weighs more than the normal range which can be in the form of bone, muscle mass or fat. However obesity directly relates to the abonormal body fat accumalation. A rough scale that can give an estimate of obesity can be calculated by Body Mass Index (BMI), which is obtained when a person's weight (in kg) is divided by height (meters). A BMI of 30 or more is genreally considered to be in the obese range, while the BMI of an overweight person is usually more than 25 (K. M. Flegal 2013). Obesity is a medical condition which leads to several other chronic disorders such as coronary heart disease, musculoskeletal diseases, diabetes and cancer.

2.1.1 Factors contributing to obesity

The most obvious and main cause of obesity is metabolic imbalance between calories consumed and calories burnt. A calorie is defined as the amount of energy needed to raise the temperature of water by one degree celcius. This, in turn, results in an excessive accumalation of body fat and with the course of time pounds of fat build up in the body. Global trends in food consumption indicate foods that have higer fat content (such as fast foods), an overall decrease in physical activity due to change in working trends and technological development, the evolution of modes of transportation and finally the increasing rate of urbanization contribute to obesity (K. M. Flegal 2013). Although genes are responsible for obesity they play a much smaller role than previously thought. Obesity related genes aggravate obesity be interacting with other environmental factors such as physical activity and diet (K. M. Flegal 2013).

2.1.2 Genetics, Obesity and the Environment

A gene-enviornment interaction is a phenomenon where the genotyoe of an individual defines the relative response or behaviour to any external environmental factor. When it comes to studying the effects of obesity there are two main categories of gene-environment interaction, namely geneotype-nutrition and genotype-physical activity interaction. In order to observe the effects of obesity on the types of interactions pair of –mono and dizygotic twins are studied for the influence of external environmental factors on obesity, besides the obvious genetic code. Through polymorphism and genetic studies on mono and dizygotic twins it has been possible to identify point mutations and genes responsible for the pathophysiology of obesity. Single genes that have been identified as regualtors of obesity are: LEP, LEPR, POMC, PSCK1, SIM1, MC3R, MC4R, CRHR1, CRHR2, SIM1 and GRP24 (C. Bouchard 2008). These genes were identified through genetic screening in individuals that were heavily obese with an early onset of obesity. Furthermore loci for obsity related genes that have a Mendelian pattern of heredity have also been identified (C. Bouchard 2008). Interaction of environment with obesity related genes was identified in a study by Bouchard, et al, 1990 where 12 pairs of twins (male) had an intake of 1000 calories surplus in their diet for a period of 6 days per week for a period of hundred days. The results of the study indicated a significant relationship between excess calories and individual responses. Furthermore the genetic variation was not random but it was observed as a significant within-pair resenblance in response. The study indicated that if other variables such as geneteic makeup and life style remain constant; even then certain individuals have a greater tandency to accumalate fat as compared to others. Also there is a likely genetic basis for the body fat storage tandency (C. e. Bouchard 1990).

2.1.3 Role of Hormones in Obesity

There has been extensive research in the role of hormones as causative of obesity. Changing levels of estrogen have been implicated in regulation of overall body weight (Blüher 2013). Two types of estrogen receptors ERα and ERβ are expressed in periperal tissues including adipose cells, where these receptors may play a role in adipose accumalation and

4

inflammation (Blüher 2013). A human polymorphism (Xbal) has been identified in the ERα gene which has a substitution of guanidine with adenine in exon-1 (Okura T 2003). A comparative study was conducted on two thousand pre-menopausal Japanese women that had an increased adipic mass and fat around the waist line and that had the polymorphism, in conrast to the women with the normal polymorphism. The results of the study indicate a relation of the polymorphism in the ERα gene with increased fat deposition in periperal adipose tissues (Okura T 2003). Another key metabolism regulating enzyme is leptin, which gives a powerful signal to the central nervous syetem. This results in a catabolic burst, involves limiting food intake and maximizing energy expenditure. Leptin is secreted by by adipose cells and its amount is directly linked to the amount of adipose tissue present in the body. The main modulator of leptin is estrogen; higher levels of estrogen during estrous and metestrous stages leads to higher catabolic sensitivity of leptin. Exogenous administration od estradiol-17β, in mice leads to in increased sensitivity to leptin as compared to controls (E. E. Calle 2004).

2.1.4 Human Microbiome and Obesity

The lower gastro-intestinal tract hosts a wide array of micro-organisms that help in the breakdown of several important nutritional components in the diet that would otherwise be indigestible. Several studies indicate slight differences in the gut microbiota of healthy adults through the sequencing and measurement of the 16S RNA expression of the bacterial population of the gut. There have been studies that indicate that miniscule variations in the gut flora are indicative of obesity and a normal body weight. In order to judge the role of microbiota in obesity, Turnbaugh, et al, (2009) characterized microbial communities housed by monozygotic and dizygotic twins by sequencing the 16S RNA gene. Although the micro-organisms of family members had almost the same core-microbiome, obesity was caused by phylum-level changes in the gut micro-organisms. Since the cause of obesity was a phylum level change in bacteria this means that bacteria from different phyla have very different metabolic pathways for degradation. The altered degradation pathways meant that different individuals will be able to degrade a single nutrient through different pathways. In conclusion deviations from the core-microbiome at the phylum level can predispose an individual towards obesity (Turnbaugh 2009).

2.1.5 Metabolism and Obesity

Circulating hormones such as insulin and steroid hormones have a profound effect on body metabolism and ultimately body weight. An increased BMI is indicative of a high level of circulating insulin and the amount of insulin increases with the increase in BMI. The main role of insulin is to balance the amount of excessive carbohydrates at the cellular level. Consumption of sugary foods lead to production of high levels of insulin a condition called hyper-insulinemia that ultimately leads to a state of insulin resistance. Obesity is attributed to hyperinsulinemia and insulin resistance (E. M. Calle 2004).

Obesity and insulin resistance as key players in weight gain have been studied across all ethnic groups and sufficient data has been collected for almost all body weights. As the BMI (fat deposition in the body) rises, so does the risk for insulin resistance and consequently of obesity. This implies that the amount of body fat present within the body has an important effect on insulin sensitivity (Kahn 2000).

2.2 Epidemiology of Obesity

A recent survey (2007-2008) by the National Health and Nutrition Examination Survey (NHANES) places the number of obese individuals at 68% who are 20 years or older (Ogden 2010). This is in contrast much higher than the survey carried out from 1988-1994 which reported 56% of adults above 20 as obese in the U.S (Crespo 2001). But this is not even half of the story since high percentages of obesity have been reported in young children (ages 2-19). The survey data on children puts the level of obese children at 17% based on a recent survey from 2007-2008 (Ogden 2010); which is much higher than the survey conducted from 1988-1994 that indicated obesity levels at 10% (Crespo 2001). Correlation studies between obesity and cancer have also seen a gradual increase in occurrences of obesity related cancers. An analytical study by Hock, et al., (2002) data from NCI Surveillance, Epidemiology, and End Results (SEER) gave an estimate that in 2007, in the U.S, 34,000 new cancer cases in men and 50,500 in women were reported that were obesity related. Recent broad survey data suggests that obesity is a leading cause of a wide range of cancers including esophagus, colon, breast, kidney and thyroid (Gilbert 2013).

3. Discussion

Obesity and eating habits account for almost 30% of the cancers in developed western countries, thus making obesity the second largest cause of cancer, after tobacco (E. E. Calle 2004). However, in developing countries the rate of cancer related to obesity is much lower and accounts for 20% of cancers (E. E. Calle 2004). Research into obesity that leads to cancer has lead to some certain facts and also some unanswered questions. In this discussion, the metabolic imbalance that lead to cancer, epidemiological studies, intervention trials, experimental studies, epidemiology and gender distribution of cancer. Furthermore, different types of cancers, which are caused by obesity, are also discussed.

3.1 The correlation of obesity and cancer: a metabolic study

There has been incresing evidence that the body weight has serious effects on the levels of circulating steroid and peptide hormones. These hormones have the capability to play a major role in cancer. A cross sectional study on obesity shows that there is an increase of insulin and Insulin like growth factor-binding protein 3 (IFGBP-3) in both males and females. Furthermore there is a decrease in IGF-1, human growth hormone and sex hormone binding protein. Also there is an increase of circulating estradiol in males and post-menopausal women under obese conditions (IARC 2011).

Perhaps one of the most noticeable increase is of insulin, in obesity. A linear increase has been observed in circulating levels of insulin which directly correlates with the BMI of the individual. Insulin is a peptide based hormone, which is secreted when blood sugar levels increase and acts to limit and control the uptake of glucose by peripheral tissues. When the body is continually subjected to large meals, over-eating and excessive consumption, the body assumes a state of chronic hyperinsulinemia. Abdominal obesity, in particular arises from hyperinsulinemia and subsequent insulin resistance (Braun 2011).

One of the mitogens that regulate energy dependent growth process are the Insulin like growth factors (IGF's). The role of IGFs is to stimulate cell growth and proliferation and down regulates apoptosis. Furthermore strong mitogenic effects on cancer cell lines have also been obseved. Pitutary growth hormone (PGH), regulates the synthesis of IGF-1 and IGFBP-3. While circulating in the blood the IGF-1 is bound to IGFBP-3. As a result of obesity, there are increased circulating levels of insulin, decreased expression of IGF binding protein (IGFBP-2 and IGFBP-3) and an overall increase of free IGF-1 in peripheral blood. However there is no net increase of IGF-1 in obesity or hyperinsulinemia, but a slight decrease in circulating IGF-1. This can be due to the negative feedback mechanism of IGF-1 on GH secretion (Braun 2011).

Elevated levels of insulin and bioavailable IGF-1 in turn up-regulate the biosynthesis and bio-availability of steroid sex hormones, that have a known part in progression of certain cancer types (Braun 2011). Furthermore a chronic state of hyperinsulinemia inhibits the production of sex hormone-binding globulin (SHBG), making the sex hormones bioavailable due to low amounts of SHBG. It is the unbound sex hormones (androgens and estrogens) that are actually bioavailable to the body and regulate cell growth, progression and function of tissues. Greater the amount of body fat in peripheral body, lower is the amount of circulating SHBG's (Braun 2011).

Besides the increased bioavailability of estrogens and androgens, the amount of fat deposits in the body also up-regualtes the levels of steroied sex hormones in men and postmenupausal women through the aromatization of androstenedione to estrone. However in postmenupausal women, there is a very low rate of synthesis of ovarian estrogen and the fat deposits are the main source of estrogen production (Braun 2011).

According to a study by IARC, (2002), reverse effects occur when an obese individual loses weight. Weight loss leads to a reduction in insulin resistance and an overall increase in the circulating levels of estrogen, insulin, glucose and SHBH. Metabolism can be improved by a net 5-10% reduction in overall body weight (IARC 2011).

3.2 Analytical Epidemiological Studies on Obesity and Cancer

During the past 30 years, studies were performed on individuals and their subsequent risk for developing cancer and the factors associated with it. The earliest studies were case-control trials that were analyzed in a retrospective fashion; such that the individuals with cancer were questioned about their diets and their eating habits were then compared with those who did not have cancer. The earlier studies were helpful in determining the role of diet and dietary components on cancer, but did not exactly describe the relationship between obesity and cancer. These reliability and accuracy of these studies was also low because they overlooked the possible smaller dietary components that may have been a causative of cancer. And since the trials were based on selected populations and not random individuals the data is also susceptible to selection bias (Cappellani 2011). Prospective studies are better at determining the cause and occurrence of cancer in relation to obesity, due to the above mentioned reasons.

3.3 Intervention Trials on Obesity and Cancer

Intervention trials regarding obesity and cancer aim to study the situation at a more realistic level, where "intervening" factors (such as dietary habits, rate of physical activity and smoking) of the population can also be accounted for. Intervention trials are more predictive and better at explaining the role of obesity in causing cancer as compared to simple observational studies. Intervention trials are carried out by assigning the participants random dietary changes or supplements, which is helpful in eliminating any misleading factors and hence gives a more reliable result as compared to simple observational studies. Intervention studies that include both obesity and cancer, testing the hypothesis is difficult because the sample size has to be very large and naturally the expenses involved are much higher. Furthermore only a single or at maximum two to three interventions can be tested, otherwise the data would become too complicated to deduce any meaningful conclusion. For example intervention trials that monitor the effect of vitamin supplements on diet and obesity, it can easily be administered. However macronutrients in each individual's diet are difficult to judge since it is difficult to blind the participants as opposed to a placebo in place of a vitamin capsule (Albanes 1995).

3.4 Experimental studies on Obesity and Cancer

A large number of studies have been conducted on incidence of cancer and its relation to obesity. Furthermore, experimental research, which was helpful in identifying the key mechanisms, that leads to cancer whilst in a state of obesity. Cell lines, animal models and humans were subjected to a molecular analysis in order to judge the disorders in hormones, metabolites and overall cellular and body metabolism turnover rates. Insulin, IGF-1 and estrogen have been identified as potential targets through experimental studies on obesity and cancer (Gilbert 2013, K. M. Flegal 2013, C. Bouchard 2008).

3.5 Cancer and Obesity in perspective of Epidemiology

Historical Overview of Studies on Obesity

Excess body weight has been implicated in mortality from various cardiovascular diseases has been well established in several epidemiological studies (Goldhaber 1997, Willett 1995, Stevens 1998). Furthermore excess body weight has been associated with an increased risk of an array of diseases including cardiovascular diseases, hypertension, glucose intolerance, osteoarthritis, type II diabetes mellitus and dyslipidemia (National Institutes of Health and National Heart Lung and Blood 2009). An overall survey of previous and recent studies on morbidity of cancer indicate that diseases and symptoms such as hypertension, dyslipidemia and diabetes emerge earlier as compared to cancer in obese individuals. However it has been difficult to associate obesity in general to specific types of cancer. Furthermore the complications that arise from obesity such as diabetes, hypertension and cardiovascular diseases leads to greater cases of mortality as compared to cancer due to obesity. Another difficulty while connecting obesity to cancer is the insufficient data on obesity as a sole cause of cancer, without including endocrinal disturbances. Therefore due to the gap in knowledge in this aspect has made obesity and cancer an active field of research. Studies on obesity and cancer have been able to identify that obesity is not accountable for just a few cancers but a large number of cancers are caused by obesity.

The IARC Report on Cancer and Obesity

A detailed assessment by the International Agency for Research on Cancer (IARC) on Obesity and Cancer was recently published as a comprehensive review of the available literature on obesity and cancer which analyzed epidemiology, clinical significance and the available experimental data on obesity and cancer (IARC 2011). Their report concluded that there was enough evidence to support that role of weight loss as a cancer preventive of colon cancer, female breast cancer in postmenopausal women, cancer of the endometrium, esophagus and kidney (IARC 2011). In the aspect of breast cancer in post menopausal women, the report concluded that the habit of avoidance of weight gain did not have a profound anti-cancer effect. For the remaining types of cancers, the IARC concluded that there was insufficient data to clearly say that avoidance of weight gain as a cancer preventive.

The conclusions presented by the IARC on cancer and obesity were based in epidemiologic studies that compared obese individuals with those that were leaner. The studies chosen however did not aim to summarize the effect of weight loss on cancer and as a cancer preventive. The reason is that there are very few individuals that are able to lose weight and successfully maintain it at a healthy weight. Therefore it becomes difficult to study the cancer-preventive effects of weight loss in larger sets of populations. This was the main reason that the IARC concluded evidence of weight loss as a cancer-preventive as insufficient and inconclusive.

The IARC also reviewed the cancer-preventive effect of weight loss in animal models that had chemically and spontaneously induced cancers of the mammary gland, pituitary, liver, adenoma, pancreas, melanoma, prostate and colon cancer. Cancers in animal models that were induced via genetic disruptions that included lymphomas were also reviewed by IARC. The basic designs of the studies were weight control through caloric restriction and the effect of cancer progression on the dietary habits. There is supporting evidence that weight loss, caloric restriction and weight loss from an obese state dramatically reduces the risk of cancer and prevents from both types of spontaneously and chemically induced cancers in animal models (Dunn 1997, Hursting 1997). Proposed mechanisms for the cancer-preventive role of weight loss are reduction of IGF-1 levels in animals subjected to caloric restriction (Dunn 1997), increased carcinogen metabolism, greater capacity to repair damaged DNA and lower oxidative stress induced DNA damage (IARC 2011).

Colorectal Cancer

Two case control cohort studies have verified the risk of cancer in relation to obesity in both men (with a relative risk of 1.5-2.0) and in women (having a relative risk of 1.2-1.5) (IARC 2011). Similar results were obtained for colon adenomas, with a greater propensity for larger adenomas (Bird 1998). There is an obvious pattern of gender difference in the occurrence of adenomas, where men are more prone to develop cancer as compared to women and the pattern has been consistent in studies across different populations. There are several theories to explain the gender biasness; one of them is that individuals with a propensity to deposit fat in central areas of the body are more prone to colon cancer as compared to individuals that have fat deposits in peripheral areas of the body. Since men usually have fat deposits in the central areas as compared to women who have fat deposits in the peripheral regions, the men are at a greater risk of developing colon cancer. A study by Giovannucci et al 1995, studied the correlation of waist circumference (measured as waist-to-hip ratio) and the development of colon cancer in the form of large adenomas in men. The study suggested a strong link between the two factors, that is waist-to-hip ratio and colon cancer (E. A. Giovannucci 1995). However there have been no significant studies performed on the correlation of waist-to-hip ratio and colon cancer in women. There was a study by Calle et al 2004 that studied the overall BMI of women and correlated it with development of colon cancer which displayed a strong relation as opposed to body fat distribution among women. Therefore the theory that body fat distribution is the explanation for the gender biasness of colon cancer is not completely justified. Another explanation for this phenomenon is that circulating estrogens (administered exogenously) in post menopausal women may lead to a weight gain but also have roles in cancer protection from adenomas (E. E. Calle 2004).

Giovannucci first proposed that it was not merely the body fat distribution but also the overall body mass index that was the determinant of risk of colon cancer. Furthermore the presence of central obesity was one of the factors that further aggravated the situation (E. A. Giovannucci 1995).

It is known that Insulin and IGF's promote growth of the mucosal cells of the colon and also promotes growth of colonic carcinoma cells *in vitro* (E. M. Calle 2004). This hypothesis has been supported with many epidemiologic studies. A high risk of colorectal cancer has been associated with elevated levels of blood glucose and insulin levels in fasting state (E. A. Giovannucci 1995). Several studies designs such as prospective cohorts (Holmes 2002) and case control studies (Campbell 2010) have concluded that an increased risk of colorectal cancer and large adenomas are directly related to increased levels of IGF-1 and low levels of IGFBP-3.

Breast cancer

Epidemiology studies designed since the 1970's have aimed at correlating the anatomical build and the risk of breast cancer and the prediction of breast cancer (IARC 2011). The early studies were helpful in establishing basic facts with the occurrence of breast cancer, such as the correlation of body size and concurrent development of breast cancer was observed in post-menopausal women who obese. However such correlation was not established in pre-menopausal obese women (IARC 2011). There is also a lower risk of developing breast cancer in pre-menopausal obese women because they have lower circulating steroid hormones and also because they may have anovulatory menstrual cycles. Furthermore obesity has a profound effect on the incidence of breast cancer, where there was a 30-50% increase in risk of breast cancer in post-menopausal women (Yang 2011). Some studies have highlighted that obesity alone is an important enough factor in post-menopausal women as a predictor of breast cancer. Furthermore weight gain post-menopause has been associated with a greater risk of cancer as opposed to BMI alone (Yang 2011, Cappellani 2011).

Breast Cancer Association Consortium summarized studies that were related to mortality rates and survival among breast cancer patients. The survey indicated that excess body weight was responsible for a higher mortality and a greater risk for a recurrence of cancer after treatment regardless of age and menopausal status. Lean women (BMI less than 20.5) have a three times greater chance of survival than obese women (BMI more than 40). Furthermore, prognosis of breast cancer in obese women is also poor since higher BMI is linked to a poorer prognosis of breast cancer, which is reliant on the number of active estrogen receptors. This makes diagnosis difficult in estrogen receptor positive tumors in stage I and II of the cancer (Yang 2011). Women with excess weight are less likely to undergo mammography screen since a higher BMI lessens the chances of self detecting a tumor. There was a significant association between BMI and breast cancer in postmenopausal women, who never used estrogen replacement therapy (ERT). The reason may be the high levels of estrogen circulating in the blood stream make it difficult to link breast cancer to BMI, regardless of the weight of the patient (Demark-Wahnefried and Jennifer A. Ligibel 2012). The observation that BMI is strongly related with occurrence of breast cancer in postmenopausal women who do not use hormone replacement therapy, may be explained by the increase in endogenous estrogen production which can aggravate the incidence of breast cancer. Furthermore, high levels of circulating endogenous estrogens and concurrently low levels of SHBG which can lead to breast cancer (Cappellani 2011).

A mechanism involving insulin and IGF's has also been described that leads to breast cancer. As discussed earlier that IGF-I is a potent stimulator of cell growth and differentiation in normal epithelial and breast cancer cells and can lead to mammary tissue hyperplasia and concurrently cancer. IGF-I receptors are found in normal breast tissue and also in breast tumors. Case-control studies and prospective cohorts have positively associated IGF-I with the occurrence of breast cancer in premenopausal women. The amount of circulating IGF binding protein 3 (IGF-BP3) and IGF-I was directly proportional to the risk of breast cancer. The incidence of IGF-I related breast cancer is higher in premenopausal women as compared to postmenopausal women (IARC 2011).

Endometrial cancer

There is an increasing amount of evidence from case-control and cohort studies that obesity and excess weight is a strong risk factor for endometrial cancer. There is also some supporting evidence on increasing BMI as a risk for endometrial cancer but not all data supports this hypothesis. A gradual increase in risk has been reported, in case of endometrial cancer in overweight or obese women, which may be two to four times greater than normal weighted women (Arem 2013).

The suggested mechanism for development of endometrial cancer is the increase in the amount of circulating estrogens with an increase in weight or in an obese condition. The risk for endometrial cancer significantly increases when postmenopausal obese women take estrogens without any hormone to balance its effects such as progesterone (Arem 2013). The bioavailability of circulating estrogens is also an important factor in progression of endometrial cancer (IARC 2011).

Kidney cancer

Obesity has shown to increase the risk for renal cell cancer by 1.5 to 2.5 times as opposed to individuals (both men and women) with a normal body weight. A large number of studies suggest a dose-response relationship as the BMI or weight increases (IARC 2011). The risk for developing renal cell cancer is greater for women as compared to men, a phenomenon which is yet unexplained. Although blood pressure is an important factor that selectively puts a pressure on the renal cells, however the development of cancer due to obesity is seems to be independent of this high blood pressure. Therefore the mechanisms for development of renal cell cancer are different than anticipated. The theory is that chronic hyperinsulinemia, which is caused due to an obese state leads to the association of BMI and renal cell cancer. This theory is supported by the studies on diabetics (that are in a chronic hyperinsulinemic state) which have a high risk of renal cell cancer (Kahn 2000, IARC 2011).

Liver and Stomach cancer

There is a limited amount of data on the association of obesity and liver and/or stomach cancer. The studies indicated an increased risk for liver cancer (20-40%) results of which were unanimous for both men and women. Furthermore the risk for developing adenocarcinoma of the gastric inner lining has also been found to be associated with obesity. Non-alcoholic fatty liver disease (NAFLD) and a more severe form steato-hepatitis (NASH) has been associated with increase in body weight and higher BMI. NASH is usually characterized by the fibrotic lesions and cirrhosis of the liver. Cirrhosis, a permanent condition in obese individuals, is also a risk factor for cancer. During the stage of liver cirrhosis the liver attempts at repairing the damage through the release of cytokines. An excess of cytokines has shown to be a causative of cancer progression in the liver (Sun 2012).

Prostate cancer

A large amount of experimental data has accumulated that disregards the relationship between obesity and the occurrence of prostate cancer. However few studies do support the theory that later stages of prostate cancer can aggravate due to obesity and may lead to death in individuals with a high body weight (E. M. Calle 2004, IARC 2011).

4. Conclusion

There is an increasing amount of data on obesity as a risk factor for a large number of cancer types. Not only obesity and overweight can lead to cancer development but obesity also has a role in poor prognosis and survival of cancer patients. Adequate steps should be taken to control obesity and it must be considered as an epidemic that is affecting all populations of the world. Furthermore, obesity and cancer monitoring, must be performed by government and related research bodies at regular intervals in order to assess the trend of obesity and cancer.

References

Albanes, Demetrius, Olli P. Heinonen, Jussi K. Huttunen, Philip R. Taylor, Jarmo Virtamo, B. K. Edwards, Jaason Haapakoski, Matti Rautalahti, A. M. Hartman, and Juni Palmgren. "Effects of alpha-tocopherol and beta-carotene supplements on cancer incidence in the Alpha-Tocopherol Beta-Carotene Cancer Prevention Study." *The American journal of clinical nutrition*, 1995: 1427S-1430S.

Arem, H., and M. L. Irwin. "Obesity and endometrial cancer survival: a systematic review." *International Journal of Obesity*, 2013: 634-639.

Bird, Cristy L., Harold D. Frankl, Eric R. Lee, and Robert W. Haile. "Obesity, weight gain, large weight changes, and adenomatous polyps of the left colon and rectum." *American journal of epidemiolog*, 1998: 670-680.

Blüher, Matthias. "Importance of estrogen receptors in adipose tissue function." *Molecular metabolism*, 2013: 130.

Bouchard, C. "Gene–environment interactions in the etiology of obesity: defining the fundamentals." *Obesity*, 2008: S5-S10.

Bouchard, Claude, et al. "The response to long-term overfeeding in identical twins." *New England Journal of Medicine*, 1990: 1477-1482.

Braun, Sandra, Keren Bitton-Worms, and Derek LeRoith. "The link between the metabolic syndrome and cancer." *International journal of biological sciences*, 2011: 1003.

Calle, Eugenia E., and Michael J. Thun. "Obesity and cancer." *Oncogene* 23, no. 38 (2004): 6365-6378.

Calle, Eugenia E., and Rudolf Kaaks. "Overweight, obesity and cancer: epidemiological evidence and proposed mechanisms." *Nature Reviews Cancer* 4, no. 8 (2004): 579-591.

Calle, Eugenia, Michael. "Obesity and cancer." *Oncogene* 23, no. 38 (2004): 6365-6378.

Campbell, Peter T., Elizabeth T. Jacobs, Cornelia M. Ulrich, Jane C. Figueiredo, Jenny N. Poynter, John R. McLaughlin, Robert W. Haile et al. "Case–control study of overweight, obesity, and colorectal cancer risk, overall and by tumor microsatellite instability status." *Journal of the National Cancer Institute*, 2010: 391-400.

Cappellani, A., M. Di Vita, A. Zanghi, A. Cavallaro, G. Piccolo, M. Veroux, M. Berretta, M. Malaguarnera, V. Canzonieri, and Menzo E. Lo. "Diet, obesity and breast cancer: an update." *Frontiers in bioscience (Scholar edition)*, 2011: 90-108.

16

Crespo, Carlos J., et al. "Television watching, energy intake, and obesity in US children: results from the third National Health and Nutrition Examination Survey, 1988-1994." *Archives of pediatrics & adolescent medicine* 155, no. 3 (2001): 360-365.

Demark-Wahnefried, Wendy, Elizabeth A. Platz,, and Cindy K. Blair, Kerry S. Courneya, Jeffrey A. Meyerhardt, Patricia A. Ganz et al Jennifer A. Ligibel. "The role of obesity in cancer survival and recurrence." *Cancer Epidemiology Biomarkers & Prevention*, 2012: 1244-1259.

Dunn, Sandra E., Frank W. Kari, John French, Joel R. Leininger, Greg Travlos, Ralph Wilson, and J. Carl Barrett. "Dietary restriction reduces insulin-like growth factor I levels, which modulates apoptosis, cell proliferation, and tumor progression in p53-deficient mice." *Cancer research*, 1997: 4667-4672.

Flegal, Katherine M., Brian K. Kit, and Barry I. Graubard. "Overweight, Obesity, and All-Cause Mortality—Reply." *JAMA* 309, no. 16 (2013): 1681-1682.

Flegal, Katherine M., et al. "Association of all-cause mortality with overweight and obesity using standard body mass index categoriesa systematic review and meta-analysisall-cause mortality using bmi categories." *JAMA* 309, no. 1 (2013): 71-82.

Gilbert, Candace A., and Joyce M. Slingerland. "Cytokines, obesity, and cancer: new insights on mechanisms linking obesity to cancer risk and progression." *Annual review of medicine* 64 (2013): 45-57.

Giovannucci, Edward, Alberto Ascherio, Eric B. Rimm, Graham A. Colditz, Meir J. Stampfer, and Walter C. Willett. "Physical activity, obesity, and risk for colon cancer and adenoma in men." *Annals of internal medicine*, 1995: 327-334.

Goldhaber, Samuel Z., Francine Grodstein, Meir J. Stampfer, JoAnn E. Manson, Graham A. Colditz, Frank E. Speizer, Walter C. Willett, and Charles H. Hennekens. "A prospective study of risk factors for pulmonary embolism in women." *JAMA*, 1997: 6.

Hock, Lynette M., James Lynch, and K. C. Balaji. "Increasing incidence of all stages of kidney cancer in the last 2 decades in the United States: an analysis of surveillance, epidemiology and end results program data." *The Journal of urology* 167, no. 1 (2002): 57-60.

Holmes, Michelle D., Michael N. Pollak, Walter C. Willett, and Susan E. Hankinson. "Dietary correlates of plasma insulin-like growth factor I and insulin-like growth factor binding protein 3 concentrations." *Cancer Epidemiology Biomarkers & Prevention*, 2002: 852-861.

Hursting, Stephen D., Susan N. Perkins, Charles C. Brown, Diana C. Haines, and James M. Phang. "Calorie restriction induces a p53-independent delay of spontaneous carcinogenesis in p53-deficient and wild-type mice." *Cancer research*, 1997: 2843.

IARC. "IARC Handbooks of Cancer Prevention." *Weight Control and Physical Activity*, 2011: 54.

Kahn, Barbara B., and Jeffrey S. Flier. "Obesity and insulin resistance." *Journal of Clinical Investigation* 106, no. 4 (2000): 473-481.

National Institutes of Health and National Heart Lung and Blood, Alberti, K. G. M. M., Robert H. Eckel, Scott M. Grundy, Paul Z. Zimmet, James I. Cleeman, Karen A. Donato, Jean-Charles Fruchart, W. Philip T. James, Catherine M. Loria, and Sidney C. Smith. "Harmonizing the Metabolic Syndrome A Joint Interim Statement of the International Diabetes Federation Task Force on Epidemiology and Prevention; National Heart, Lung, and Blood Institute; American Heart Association; World Heart Federation; International A." *Circulation*, 2009: 1640-1645.

Ogden, Cynthia L., et a. "Prevalence of high body mass index in US children and adolescents, 2007-2008." *Jama* 30, no. 3 (2010): 242-249.

Okura T, Koda M, Ando F, Niino N, Tanaka M, Shimokata H. "Association of the mitochondrial DNA 15497G/A polymorphism with obesity in a middle-aged and elderly Japanese population." *Human Genetics*, 2003: 432-436.

Stevens, June, Michael W. Plankey, David F. Williamson, Michael J. Thun, Philip F. Rust, Yuko Palesch, and Patrick M. O'Neil. "The Body Mass Index-Mortality Relationship in White and African American Women." *Obesity research*, 1998: 268-277.

Sun, Beicheng, and Michael Karin. "Obesity, inflammation, and liver cancer." *Journal of hepatolog*, 2012: 704-713.

Turnbaugh, Peter J., Micah Hamady, Tanya Yatsunenko, Brandi L. Cantarel, Alexis Duncan, Ruth E. Ley, Mitchell L. Sogin et al. "A core gut microbiome in obese and lean twins." *Nature* 457, no. 7228 (2009): 480-484.

Willett, Walter C., JoAnn E. Manson, Meir J. Stampfer, Graham A. Colditz, Bernard Rosner, Frank E. Speizer, and Charles H. Hennekens. "Weight, weight change, and coronary heart disease in women: risk within the'normal'weight range." *Jama*, 1995: 461-465.

Yang, Xiaohong R., Jenny Chang-Claude, Ellen L. Goode, Fergus J. Couch, Heli Nevanlinna, Roger L. Milne, Mia Gaudet et al. "Associations of breast cancer risk factors with tumor subtypes: a pooled analysis from the Breast Cancer Association Consortium studies." *Journal of the National Cancer Institute*, 2011: 250-263.